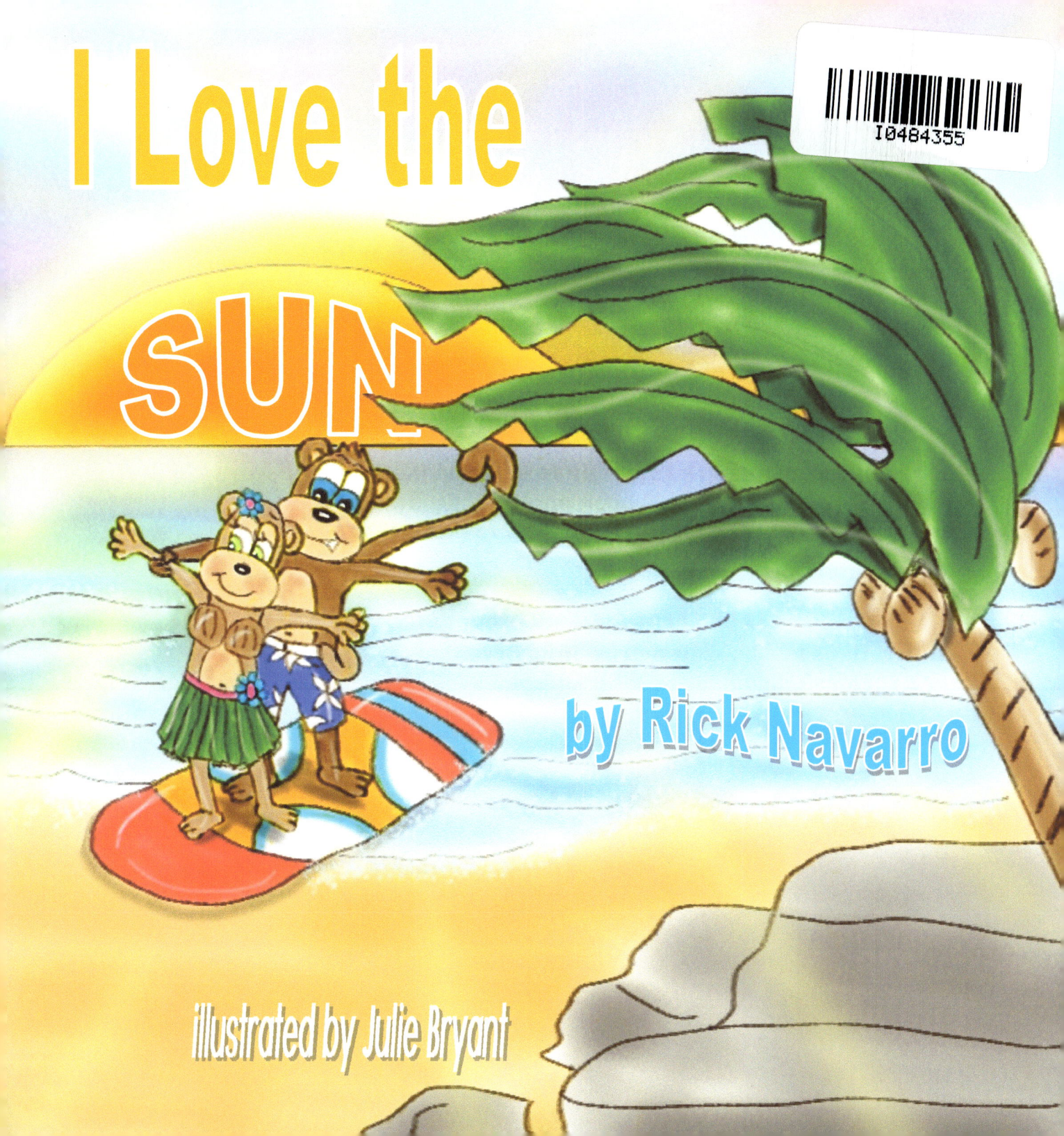

I Love the SUN

by Rick Navarro

illustrated by Julie Bryant

HERBIE

Illustrations and book design by Julie Bryant of *sweetartdesign.com*

Grapevine, Texas

2014

JENNY

DEDICATION

This book is dedicated to the children of the world in the hope that they will love science, nature and God for the benefit of all creatures, large and small.

I love the sun
Up in the sky.

It makes me happy
I wonder why?

The sun sends love
From far away
To wake me up
At the start of day.

It warms me up
When cold winds blow.

And helps me live
From head to toe.

Sunshine's the fastest
In the whole wide world.
Faster than any boy or girl,

Rabbit, cheetah, dog or cat
Sunshine is the fastest yet!

I love the sun
Up in the sky.
It makes me happy
I wonder why?

The sun warms the ocean
To make the clouds
That bring the rain
And thunder LOUD.

When the rain comes down
And the sun is there,
It makes a rainbow
In the air.

The sun warms the Earth
And sends up heat
To help the eagles
Fly high and neat!

I love the sun
Up in the sky.
It makes me happy
I wonder why?

Without the sun
There'd be no life.
I'm so glad
It shines so bright.

My food all comes
From the sun,
To fill my tummy
With lots of YUMM!

Apples, peaches, carrots and pears
All need the sun to be there.

Even grass the cows all eat,
Need the sun
To grow green and sweet.

The cows eat grass
And then they moo,
To make more milk
For me and you.

I love the sun
Up in the sky.
It makes me happy
I wonder why?

When sunshine comes to my home,
I can use it all alone.
The sun can warm or cool my house,
Or dry my shirt or my blouse.

It makes hot water
for a bath,

Or TV cartoons that
make me laugh.

On a sunny, sunny day
We go to the beach
To swim and play.

Yet too much sun is not too good.

It can burn my skin,

Like mom said it would.

So, we add some screen,

And shade our eyes

So, we can play and play

outside.

When the sun goes down.
It's time to sleep
And rest my head,
My arms, my feet.

I love the sun
Up in the sky.
And now we know
Exactly why!

More Books by Rick Navarro:

Resurrection of the Blue Planet
The Plan to Reduce Global Warming

Richard Navarro, Ph.D.
An Inhabitant of Planet Earth

Watch for More Books:

I Love the Sea

Maggie, the Lonely Tree